# ENERGY IN
# EUROPE

Mark Smalley

**EUROPE**

Energy in Europe
Europe and the Environment
Europe and its History
Farming in Europe
Industry in Europe
Living in Europe
Tourism in Europe
Transport in Europe

Editor: Janet De Saulles
Series Design: Bridgewater Design
Book Design: Jackie Berry

First published in 1991 by Wayland (Publishers) Ltd.,
61, Western Road, Hove, BN3 1JD, England

British Library Cataloguing in Publication Data

Smalley, Mark
   Energy in Europe. — (Europe)
   I. Title   II. Series
   333.79094

ISBN 0 7502 0253 X

Typeset by Dorchester Typesetting Group Ltd
Printed in Italy by G. Canale C.S.p.A., Turin
Bound in France by A.G.M.

**ACKNOWLEDGEMENTS**

BP 29; Chapel Studios 8 (top); Eye Ubiquitous (Peter Blake) 4 (top), (John Hulme) 7, 12 (top), (Mark Newman) 39; Greenpeace (Leitinger) 34; Rex Features 11, (Anastasselis) 15 (bottom); Science Photo Library 18, (Simon Fraser) 24, (Martin Bond) 36; Shell 15 (top); Tony Stone 3, 8 (bottom); (JW Sherriff) 23, 25; Topham 6, 16, 17, 20 (bottom), 22, 35; WPL 12 (bottom), 19, 20 (top), 29; Zefa (Jed Sharp) 4 (bottom), 10, 14 (bottom), 32, 37, 38, (Starfoto) 41, (Eichhorn – Z) 43, (Streichan) 44. Cover artwork and all interior artwork by Malcolm Walker.

# Contents

Introduction .................................................................4

What is energy? .........................................................8

Energy in Western Europe .....................................11

Energy in Eastern Europe ......................................16

The fossil fuels .........................................................19

Coal ...........................................................................21

Oil ...............................................................................26

Gas ..............................................................................28

Nuclear power .........................................................31

Energy conservation ..............................................36

Renewable energy ...................................................38

The way forward .....................................................44

Glossary .....................................................................46

Books to read ...........................................................47

Further information ................................................47

Index ..........................................................................48

# Introduction

This book looks at the role of energy in European society. Energy is necessary for all kinds of activities to take place. As humans we need energy to breathe, and to keep the blood flowing around our bodies. We use energy walking down the street, when riding a bike and when sleeping. But energy is also required to boil a kettle, power a car, or produce this book, just as a plant needs energy in order to grow.

Above **We need energy for all our actions, from the simplest of movements to the most demanding of activities.**

Left **Gas provides the high temperatures necessary for glass blowing in this furnace.**

4

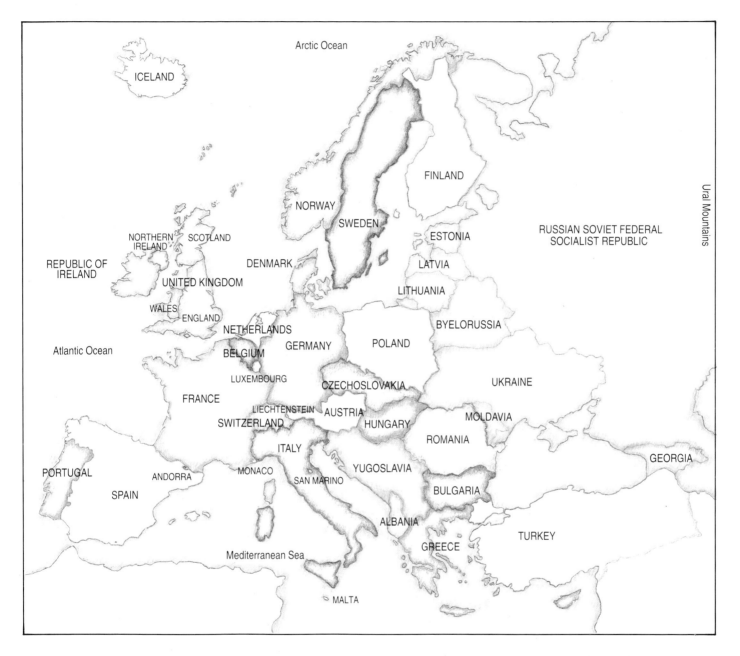

**The countries of Europe.**

We think of energy in so many different ways, because it comes in varied forms. Yet without it we could do nothing. Energy is simply the power of doing work.

All sources of energy on earth can be traced back to the sun. The sun's warmth enabled plants to grow millions of years ago, before they became fossilized and formed coal, gas and oil deposits under the earth's surface. Today, the industries and economies of Europe rely very heavily on these three forms of energy: the fossil fuels. Whilst food is our fuel, coal, gas and petrol are the fuels for cars, boilers and furnaces. Without these forms of energy, Europe would not have such a high standard of living.

Such a standard of living, however, can only be maintained as long as these fuels are around, or if new, alternative forms of energy become more widely used. This book looks at how European countries have organized their energy supplies and consumption up until now, and tries to assess patterns for the future.

## EUROPE TODAY

It is not easy to define Europe, because it can be looked at in so many different ways. What you consider to be Europe can depend on the particular country in which you live.

Some people think of Europe as just the twelve members of the European Community (EC). These EC countries are part of Western Europe. There are, however, also Western European countries which are not members of the EC, such as Austria and Switzerland, and the Scandinavian countries of Norway, Sweden and Finland.

Yet Europe extends even beyond these countries. If you were to ask people in the former Communist countries of Eastern Europe what they thought they were, or to ask people in the Baltic republics of the USSR, they would tell you that they were Europeans.

For the purposes of this book, Europe is defined as the continental landmass extending from the Atlantic Ocean in the west to the Ural Mountains of the USSR in the east. It is this range of mountains that separates Europe from Asia.

The lorry in this poster represents the American financial aid which helped rebuild the economies of Western Europe after the Second World War.

## EUROPE AFTER THE SECOND WORLD WAR

The end of the Second World War in 1945 saw the continent of Europe devastated. The destruction of six years of war lay behind: ahead lay the huge task of

On 9 November 1989 the divided city of Berlin was opened up. For the first time in years Germans were able to cross freely between East and West.

rebuilding cities and industries. Coal-mines which had been shut down during the war had to be reopened. Without energy supplies nothing could have been done.

During the war the USA and the USSR had been allies. Once peace was established, however, the rivalry between their two very different political and economic systems of capitalism and Communism became clear.

The USSR had liberated East Germany, Poland, Hungary, Czechoslovakia, Bulgaria and Romania from German occupation at the end of the war. With their new Communist governments, these Eastern European countries looked to Moscow for support.

While Moscow was helping Eastern Europe, the USA poured in financial aid to the Western European democracies under a scheme called the Marshall Plan. This was aimed at speeding up economic recovery, and at resisting the perceived threat of Communism from the USSR. In Winston Churchill's words, Europe was split by an 'Iron Curtain'. The increasing hostility between the two superpowers was called the Cold War. It lasted for 40 years, ending in autumn 1989 when that powerful symbol of the division of Europe, the Berlin Wall, came crashing down.

During the Cold War, Eastern and Western European societies had followed very different paths. Today, while Eastern European countries are struggling to gain national independence, the twelve Western nations of the EC are working all the time towards greater unity. The collapse of the Communist government in East Germany led to East and West Germany reuniting in October 1990, and to the East part of Germany joining the EC at the same date. Meanwhile, Austria, Cyprus, Malta and Turkey have formally applied to become members, and Finland, Iceland, Norway, Sweden and Switzerland are also thought to be considering applying. The face of Europe is indeed changing.

## What is energy?

Energy can take many forms. Mechanical, electrical, magnetic, chemical and nuclear energy all provide the power of doing work.

Above **Coal being delivered to a house in an English village.**

A moving object has energy, and it is called kinetic energy. Even a stationary object can possess energy, because of its position. Imagine a raised ____ When it falls it has kinetic ___ because of its movement, but until it does so, that energy remains stored. Such stored energy is called 'potential energy'. This type of energy is important in hydroelectric power.

Primary energy is the resource as it commonly occurs in nature. This might be in the form of a fossil fuel, such as coal or crude oil, or in the form of natural gas. Wind, solar

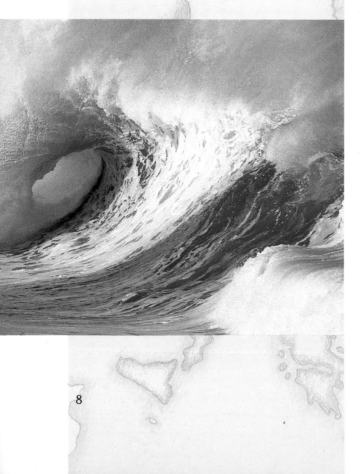

Left **The primary energy contained by the sea can be used to generate large quantities of electricity. As oil resources run out, Europe will have to do more research into renewable forms of energy.**

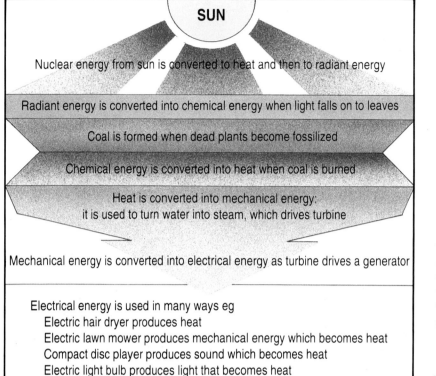

SUN

Nuclear energy from sun is converted to heat and then to radiant energy

Radiant energy is converted into chemical energy when light falls on to leaves

Coal is formed when dead plants become fossilized

Chemical energy is converted into heat when coal is burned

Heat is converted into mechanical energy;
it is used to turn water into steam, which drives turbine

Mechanical energy is converted into electrical energy as turbine drives a generator

Electrical energy is used in many ways eg
  Electric hair dryer produces heat
  Electric lawn mower produces mechanical energy which becomes heat
  Compact disc player produces sound which becomes heat
  Electric light bulb produces light that becomes heat

Above **This diagram shows how all our energy derives from the sun.**

Right **A diagram showing the proportions of fuel used in an average household.**

and wave power are also forms of primary energy.

Secondary energy results from the processing of the primary resource, for example when oil is refined t̲ ̲ ̲ ̲e petrol, or when coal or w̲ ̲ d̲-p̲o̲w̲e̲r̲ are used to generate electricity. In these cases, petrol and electricity are the secondary energy sources.

Delivered energy is what is delivered to our homes and to industry. Gas, petrol, coal and electricity are examples of this type of energy.

Useful energy is the amount of delivered energy that actually provides a useful service, such as keeping people warm, cooking food, or supplying light or mechanical work. The difference between delivered and useful energy is the amount of energy lost. This can be because of badly insulated buildings or inefficient appliances.

## ELECTRICITY

Electricity is generated by a combination of means. Nuclear power provides electricity all the time. Coal-, oil- and gas-fired power-stations, as well as hydroelectric power, can be turned on and off according to what the demand is at the time.

It would be quite wrong to think

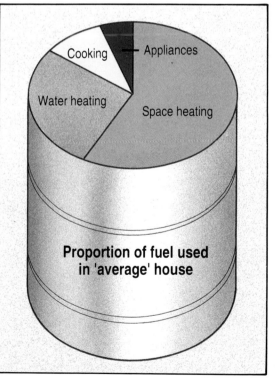

Cooking · Appliances

Water heating · Space heating

**Proportion of fuel used in 'average' house**

statistics in this book about the West of Europe than the East. This is because up until recent years, unlike Eastern European countries, the EC and other Western European countries made their facts and figures readily available.

As with statistics about the past, so should statistics about the future be treated with caution. In the 1960s it was said that within thirty years nuclear power would be producing at least 50 per cent of Europe's electricity. The figure in 1990 was, in fact, only around 10 per cent.

Below **A diagram showing the quantities of energy produced and consumed in the EC.**

that electricity is the only form of energy we use. Only 10 per cent of delivered energy actually takes the form of electrical energy, needed, for example, for things like lighting, TV, motors and appliances. 25 per cent of delivered energy is liquid transport fuel, and 65 per cent is used to supply heat, derived from petrol, oil, gas, coal and electricity.

## WHAT ENERGY IS USED FOR

Energy is used in different areas of society. As a rough average, in Western Europe industry uses 35 per cent of delivered energy, households 30 per cent, transport 20 per cent, and commerce and services 15 per cent.

In general, there are far more

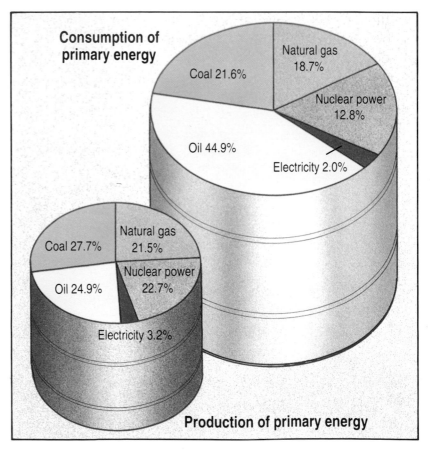

Consumption of primary energy

Natural gas 18.7%

Coal 21.6%

Nuclear power 12.8%

Oil 44.9%

Electricity 2.0%

Coal 27.7%

Natural gas 21.5%

Nuclear power 22.7%

Oil 24.9%

Electricity 3.2%

**Production of primary energy**

# Energy in Western Europe

An important step on the road to post-war peace and reconstruction in Western Europe was the foundation in 1951 of the European Coal and Steel Community (ECSC). The region of Alsace-Lorraine, important because of its coal reserves, had long been a source of conflict between Germany and France. The ECSC attempted to co-ordinate joint action over coal and steel production between these old enemies in order to prevent further conflict. Italy, Belgium, the Netherlands and Luxembourg also joined. Together they became known as 'the Six'.

Co-operation over energy policy led to the founding of the European Economic Community (EEC) in 1958 by 'the Six'. One of its foundation stones was the European Atomic Energy Community (EURATOM) which encouraged co-operation over the safe development of nuclear power between EC countries.

The proposals of Paul Henri Spaak, former Belgian Foreign Minister, formed the basis of the European Economic Community and the European Atomic Energy Commission, both founded in 1958.

Coal supplied 85 per cent of the total energy requirements of Western European countries after the war. Post-war reconstruction was so rapid that primary energy consumption doubled between 1950-73. Demand for energy so outstripped supply that the British, German, Belgian, French and Spanish coalfields were not producing enough to satisfy the needs of industry. Population growth and higher standards of living also increased energy demand.

Oil, which had been newly discovered in the Middle East,

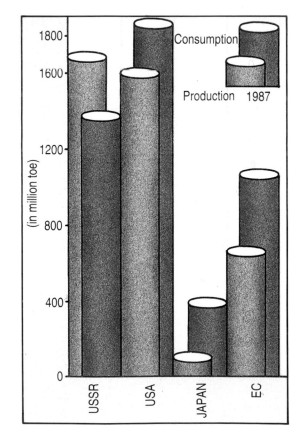

Top **A huge oil terminal in Rotterdam, the Netherlands. Here, tankers unload crude oil which is then refined and distributed by pipeline.**

Right **How the production and consumption of energy in the EC compares with the rest of the world.**

Left **Much of Europe needed rebuilding after the Second World War. Here, British Prime Minister Winston Churchill visits a bomb-damaged street in London.**

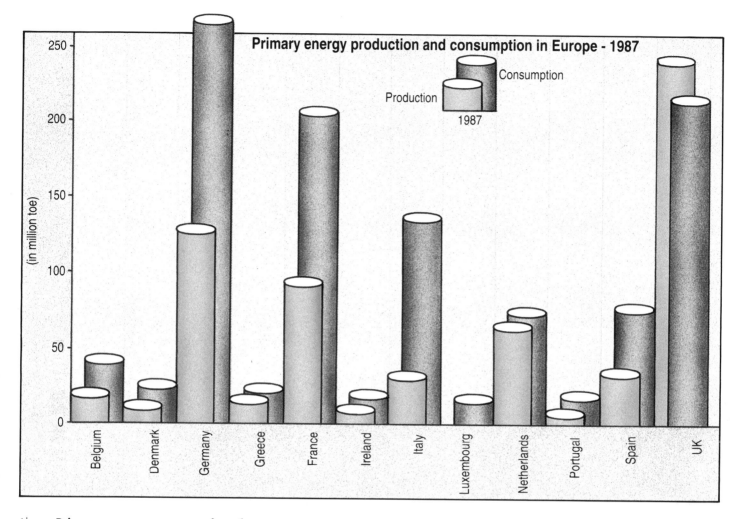

**Primary energy production and consumption in Europe - 1987**

Production Consumption

1987

(in million toe)

Belgium | Denmark | Germany | Greece | France | Ireland | Italy | Luxembourg | Netherlands | Portugal | Spain | UK

Above **Primary energy production and consumption in Europe (figures from 1987).**

proved to be one answer. By 1973, instead of being nearly self-sufficient in producing their own energy, EC countries imported 60 per cent of it. The transfer from coal to oil led to a decline in traditional heavy industry located close to the coal-mines, such as iron and steel manufacturers. Instead, the oil terminals became a focus for new industries such as petrochemicals, obtained from the refining of crude oil. The terminals are close to the major sea ports

such as Hamburg, Rotterdam and Marseilles which import crude oil.

While international oil prices remained cheap there were no problems. But the first oil crisis of 1973 was a major turning point in European and international energy policy. The price of oil quadrupled within a year. This affected industry very severely. In response, Western European countries sought to substitute their reliance on oil by broadening their use of other energy resources, such as gas, coal and nuclear energy. Serious research into renewable energy such as solar and tidal power was also undertaken.

Another effect of the oil crisis was to conserve energy, by using less of it but more efficiently, in industry, public buildings and the home.

The oil crisis affected different countries in different ways. France and Belgium rapidly developed nuclear power, while Britain, the Netherlands and Norway turned to their own newly discovered reserves of North Sea oil and gas. Some countries with very limited domestic energy reserves, such as Luxembourg, Portugal and Italy, had no choice but to depend heavily on imported energy.

Differences in the distribution of energy reserves between member countries, and the effects of the oil crises have led the EC to develop a common energy policy. On 1 January 1993 the EC becomes a single market, without any internal trade barriers between member countries. In the same way, the Single Energy Market (SEM) will one day bring to an end individual countries making their own energy policy. The idea is that the EC can then plan its energy needs as a community.

Opposite page (top) **Oil platforms in the North Sea will continue to pipe oil ashore to Britain, the Netherlands and Norway for at least another 50 years.**

Opposite page (bottom) **EC committees such as this one are held to discuss energy policies.**

Left **A huge oil refinery.**

The creation of a single market for energy within the EC will benefit France, which has surplus electricity to sell to neighbouring countries, generated by its huge nuclear power industry. However, it is also likely that the SEM will bring about the closure of coal-mines in Belgium, the Netherlands and France. This is because, compared to cheaply imported coal, it is expensive to extract. Even though the EC could still provide all the coal it needs, it imports 38 per cent of its coal from Poland, the USA, India and Australia.

## Energy in Eastern Europe

Under Communist governments, the countries of Eastern Europe sealed off their economies from international trade, and became closely linked to Soviet trade. In 1949 the Council for Mutual Economic Assistance (COMECON) was founded. This association promoted trade between socialist countries, and sought to increase industrial production. This placed great demands on Eastern Europe's energy resources.

In Western Europe the supply of goods was set by public demand. By contrast, in the USSR the long-term goals of industrial expansion and self-sufficiency were to be achieved by central planning or government command. In the same way, each Eastern European country adopted a Five Year Plan which set the priorities for industrial and agricultural production. One Year Plans then put them into action.

As part of the Five Year Plans, central government would instruct coal-mines and industries to meet particular targets. Everything was

Above **In 1990 the USSR stopped supplying oil to Lithuania in an attempt to halt the economy and undermine the independence movement. Members of the public had to queue for hours to get just a few litres of petrol.**

Opposite page **Copsa Mica in Romania has the reputation of being the most polluted town in Europe.**

geared towards achieving them. The system frequently led to shortages of some products, such as essential food items, and to surpluses of others. When shortages occurred, it was the people who suffered. In Romania, for example, if energy production did not meet the goals set, electricity supply was rationed to towns and cities.

The post-war industrial expansion of countries such as Poland and Czechoslovakia relied almost entirely upon energy produced from burning lignite, or brown coal. This type of coal has a high sulphur content and widespread use of it led to heavy pollution in and around industrial centres, such as Leipzig in former East Germany. In 1984 Poland obtained 97.5 per cent of its electricity from burning lignite.

The Eastern bloc countries bought their oil and gas from the USSR. The USSR produces one-fifth of the world's oil, and more than one-third of the world's gas. It is self-sufficient in both. For this reason, and because their economies were closed to the world market, the 1973 oil crisis did not affect Eastern Europe as much as it did the West.

Long before the 1980s, it was clear that the Communist system of centralized economic planning was not working. It was difficult for the Soviet authorities to prevent

**Pollution from a lignite-fired power-station in Bitterfeld, former East Germany.**

people envying the higher standard of living in Western Europe. Nor could the system adjust to new developments in high-technology and computer-based industries.

The Soviet economy had become too weak for the USSR to remain a superpower. The constant threat of war with the West and the continual military expansion during the Cold War took its toll. Unable to control independence movements in Eastern Europe, President Gorbachev's policy of *perestroika* (reform), opened the way for the desruction of the Berlin Wall in autumn 1989. The

momentum continued when in 1990 Communist governments of Eastern Europe were defeated in elections.

Freedom has its drawbacks for the Eastern European countries. As they race to adopt the Western European model of democratic government, and a market rather than a planned economy, they are no longer protected by their old closed system in COMECON. In order to attract Western European investment, their industries will have to become as competitive and efficient as those, for example, in what used to be West Germany. The changes, so far, have led to unemployment, and the closure of many factories. Consequently, energy demands have fallen.

Another important factor is that Western European investors are going to demand that the same high standards of environmental protection apply in the East as well as the West. This will mean a reduction in the use of brown coal for energy production. A wider use of other less polluting sources of energy, such as gas, will also take place.

The updating of Eastern European industry during the 1990s means that energy demands will probably be lower than before. This is because the more efficient industrial processes replacing the old ones will produce more goods for less energy.

# The fossil fuels

There are two main ways of extracting coal from the ground: deep and opencast mining. The first system involves the sinking of a shaft, from the surface down to the coal seams. As late as the 1920s European miners still used to hew the coal with a pick and shovel. Today, most deep mines use a longwall cutter, a machine which cuts along the length of the coal seam. The coal is then removed by conveyor belt and trains.

Opencast mining can be used if coal is discovered close to the surface. Huge earth-moving equipment removes the layers of soil before the coal is scraped away. It is a large-scale operation, and more common in Poland and the USSR than in Western Europe.

Gas was first discovered on land at Groningen in the Netherlands in 1959. The discovery of more onshore gas fields in Germany, Denmark and Britain led geologists to hope that they would find oil and gas under the North Sea.

**Opencast mining is a huge operation involving large machinery such as this German excavator.**

Right **Oil workers check the equipment at the wellhead during the drilling of a new well.**

Below **The explosion on the Piper Alpha oil platform in 1988 showed how important it is to check for high safety standards.**

Test drillings soon proved the geologists right, and oil and gas were found. The North Sea consists of two major basins. The gas fields are in the shallower southern basin, while the oil fields are in the deeper northern basin. The territory has been divided up between Britain, Norway, Denmark, Germany and the Netherlands.

Locating oil or gas is very difficult, and often hit-or-miss. 75 wells were drilled in the northern part of the North Sea before the Norwegian Ekofisk oilfield was discovered in 1969.

Developing the North Sea oil and gas fields has been a huge technological challenge, made harder by the depth of the sea, and its roughness in winter. Oil platforms are built 35 m above the sea in order to avoid high waves. Once test drillings reveal where the oil is, it is piped ashore along the sea-bed to the refinery. Workers in the British section work shifts of ten days on and ten days off, and are flown by helicopter from Inverness to the oil rigs. While there are rigorous safety regulations, when an accident does happen many lives can be lost in nightmarish conditions, as in the Piper Alpha disaster of 1988. Every precaution should be taken to make sure that similar disasters do not take place in the future.

# Coal

The Romans understood the importance of coal as a source of energy when they called it 'the stone that burned'. Coal was first found in Europe along the coast in parts of Britain, and by the thirteenth century coal trading between Britain and France was established. As the ancient forests of Europe were pushed back by the population increase and the expansion of farming, coal was used as a substitute for firewood.

Coal has a much higher calorific value than wood. This means it produces greater heat when equal quantities of both are burnt. The importance of this fact became clear in Britain during the Industrial Revolution in the eighteenth century.

Coal changed the face of north-western Europe. It brought to an end the agricultural society of the Middle Ages and hastened in today's industrial society. New factories and industrial towns grew up where coal deposits were found, first in the north of England, then in northern France, Belgium and Germany. Peasants left their life in the countryside to come and seek regular work in the new factories. Out of this change there arose a new class structure in society, based on the relationship between the factory owners and the industrial workers.

Isaac Watt, the Scottish inventor, constructed the first practical steam-engine in 1769. The steam was produced by the burning of coal and the engine was soon used to power the machinery in ironworks and cotton mills. The invention of the steam train and steamship followed. The Industrial Revolution did not only change the structure of society in Europe, but also revolutionized every form of transport.

# Coal in Wales

In 1900 South Wales was the largest coal-producing area in the world. At the heart of the region were the coalfields of the Rhondda Valley. Within 60 years the valley's population increased from 950 in 1860 to 160,000 in 1920, and 70 per cent of these people were directly involved in the coal industry.

Today's decrease in the number of coal-mines in the Rhondda shows the decline in the importance of coal in Wales. In 1913 there were 53 collieries in the Rhondda valley. By 1947 there were only 12. Maerdy, the last, was closed in 1990. The Welsh coal industry now employs less than 5,000 people, compared to 232,000 in 1913. There are only 6 deep mines left in the whole of Wales.

This decline was a result of the high cost of producing the coal from the deep mines. It became cheaper to import coal by ship from abroad. After the Second World War, Britain changed its energy policy and began to rely far more on oil than coal.

The National Union of Mineworkers began a strike in 1984 to oppose the closure of coal-mines. Mining communities throughout Britain struggled to keep their pits open, but after a year it was clear they were not going to succeed.

One miner from Maerdy talked about the effect of the pit's closure on the life of his community: 'We can't quite see where we are going – if anywhere. More than half the people here between the ages of 18 and 24 have no prospect of being employed. There's nothing for the youngsters and they just hang around the corners. Everyone has this feeling of being at the end of the line.'

The old South Wales mining valleys face a troubled future. Social problems such as unemployment are accompanied by a high death rate, particularly from cancer, heart attacks and lung infections, among retired miners. Some people hope that the 30 new Japanese companies which now employ 7,000 people in the valleys will provide ex-miners with a fresh chance of work.

**Pithead gear standing idle at the old Welsh mines of the Rhondda Valley.**

## COAL IN EUROPE

The wealth of the economies of Britain, France, Belgium and Germany was originally based on coal. In fact there is an old saying that 'Britain is built on coal'.

Some of those countries which lacked coal reserves, such as Portugal, Sweden or Finland, did not industrialize until this century, when it was discovered they could generate power by using imported oil or by hydroelectric means.

Although the Second World War severely disrupted coal production in Europe, it soon regained and then went beyond pre-war figures. In 1955 West Germany, France, Belgium, the Netherlands, Spain and Britain were producing 30 per cent of the world's output of coal. However, even this was not enough to meet the ever increasing demands of their expanding, energy-hungry economies. Extra supplies of coal were being imported from the USA. In a rapid change of energy policy, Europeans began to import cheap and plentiful oil from the USA and the Middle East. The increase in energy demand was met by oil, so much so that imports of oil doubled between 1950-55.

Coal had lost its status: it had been displaced by oil as the most important source of energy in Western Europe. This change in policy resulted in hardship and unemployment for mining communities as far apart as Spain and Germany. Producing coal was often the sole reason for such communities. Despite government retraining schemes for ex-miners, complex social problems still exist.

**The pithead winding gear, used to propel the lifts, is a symbol of coal mining.**

**Spruce trees in Czechoslavakia suffering from the effects of acid rain.**

## COAL AND THE ENVIRONMENT

As concern about environmental issues has increased across the whole of Europe, coal has proved to be a major source of pollution. The smoke and ash released from the burning of coal contains twice as much carbon dioxide ($CO_2$) as does that from the burning of gas. Scientists are increasingly sure that $CO_2$ contributes to the 'greenhouse gases' responsible for the increase in global temperatures.

The greenhouse effect is a global issue. Because Europe was at the heart of the Industrial Revolution, it has been burning fossil fuels longer than any other part of the world. It has, therefore, a major part to play in reducing emissions of greenhouse gases.

Lignite, or other low quality coal, contains a lot of impurities such as sulphur. These impurities are released into the atmosphere upon burning. The sulphur gas mixes with water droplets in the clouds to form weak sulphuric acid, and falls to the ground as 'acid rain'. Lakes and forests in Germany and the Scandinavian countries have been severely affected by this form of pollution.

In Eastern European countries such as Poland, Czechoslovakia and the former East Germany, coal remains the single most important source of energy production. Much of the coal produced in these places is low-grade lignite, and therefore plays a large part in the release of many tonnes of sulphur into the atmosphere.

## COAL AND THE FUTURE

Coal faces a difficult future. On the one hand, while world supplies of oil are running out, Britain estimates that it still has sufficient coal reserves to last another 300 years. On the other hand, because of the strong links between coal and environmental pollution, it would make sense to turn to non-polluting and renewable forms of energy such as solar and wind power. However, development of the renewables is not advanced enough for them to play a major role in energy production at the moment. One idea is that while they are being developed, coal would offer a bridge to a future form of energy production.

The coal industry itself has researched new ways of limiting pollution. Flue gas desulphurization filters installed in the chimneys of coal burning power-stations can remove 90 per cent of sulphur responsible for acid rain. However, the technique is expensive, and not all European countries have adopted it.

The EC has agreed not to increase $CO_2$ emissions after the year 2000 beyond their 1988 levels. However, as limited as this decision is, it still lacks joint agreement, since Britain is asking for another five years in which to meet the requirement.

As the Eastern European countries begin to expand their economies, they will be placing ever greater demands on their energy resources, and therefore increasing levels of pollution. The European Community is keen to encourage their use of anti-pollution technology.

**While coal is still being burnt in Europe, it is important to make greater use of anti-pollution technology.**

# Oil

A huge range of products, from petrol to plastics and textiles, are made from crude oil. Its adaptability makes it very important to Europe's industrial societies. Oil supports the high standard of living in Western Europe, and is vital to the expansion of the Eastern European and Soviet economies.

Crude oil is a thick, dark liquid made up of different kinds of oils. Where the reserves are under the sea, oil is brought to shore either by pipeline or tanker. It then has to be refined.

During the 1950s Europe was nearly self-sufficient in supplying its own energy needs. Coal provided 85 per cent of all primary energy. However, the rapid growth of Western European economies led to a huge increase in demand for energy that coal could not meet.

By the 1960s oil was in cheap and plentiful supply from the Middle Eastern oil-producing countries such as Saudi Arabia. This led to the situation where it was oil that fuelled the expansion of the post-war economies. No longer self-sufficient in supplying their own energy, the countries of Europe became reliant on imported oil. This was not a problem while regular supplies were maintained. But when shortages occurred, the problems became clear. Europe could not control its energy supplies, and relied on the good-will of the producers.

The first oil crisis in 1973 came as a big shock to European economies. The price of oil quadrupled because of shortages brought about by the war between

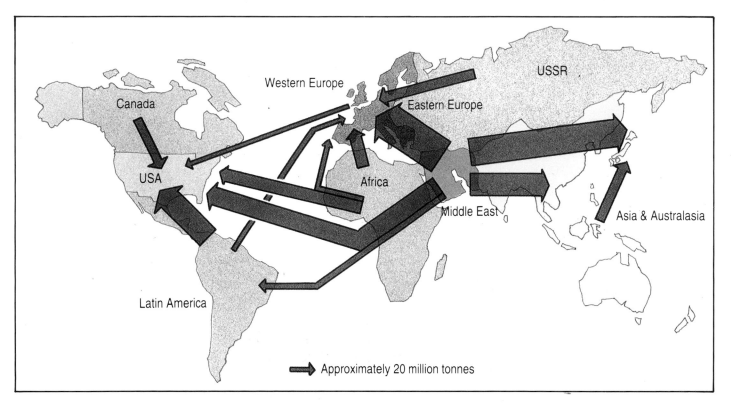

Approximately 20 million tonnes

Israel and Egypt. Without oil European industry was starved of energy. Petrol was rationed, and in the Netherlands the use of cars was banned on Sundays. It brought home the message of just how dependent the EC was on foreign supplies, and that fossil fuels would one day run out.

The same message was reinforced by the second oil crisis of 1979 when supplies were affected by the Iranian revolution and oil prices doubled. The Gulf War of 1991 showed the world the implications of being too dependent on Middle Eastern oil supplies.

The huge oil price rise of 1973 took such a big bite out of national budgets that economic activity fell throughout Europe. It led to a recession and high unemployment during the early 1980s, and it was not until 1987 that energy demand outgrew that of 1980.

The oil crisis made it urgent for Europe to find its own supplies. It led to the discovery of North Sea oil, which has benefited Britain and the Netherlands more than anyone else. However, despite its attempts to reduce its dependency on oil imports, the European Community still obtained 45 per cent of its primary energy from oil in 1987, 84 per cent of which was imported.

**A map of the world showing where oil is produced and sold.**

## Gas

Increasing concern about the environment has led to greater use of gas. Made of methane (a hydrocarbon), it is non-poisonous, and the cleanest burning of all the fossil fuels. This means it is smokeless, leaves no ash, and does not produce the poisonous carbon monoxide gas that comes from burning coal. With today's fears about global warming and the greenhouse effect, gas only releases half as much carbon dioxide as coal, or two-thirds that of oil, for the same amount of energy. Gas is commonly used in the home for central heating and for cooking. It is particularly important in the baking and glass-making industries, as well as in the preparation of nitrogen-based agricultural fertilizers.

Because of North Sea gas, Britain is now self-sufficient in natural gas, with reserves which should last until 2020. The Netherlands exports gas to neighbouring EC countries, but plans to halt its exports in 2010 in order to conserve its own supply source.

If North Sea gas was the only gas reserve in Europe then the situation would look bleak. However, the USSR, Norway and Algeria are major gas exporters, and together provide 65 per cent of the EC's needs.

Soviet gas comes from Siberia, and is piped thousands of kilometres to its customers in Eastern and Western Europe. As a clean burning fuel, gas offers one way for countries such as Poland and Czechoslovakia to lessen their dependence on brown coal. In coming years, as their economies expand and their demand for energy increases, gas will probably become a much more important energy source in Eastern Europe.

Because of the rapid changes in Soviet society since *perestroika* and the unpredictability of the future, Poland and Czechoslovakia are unwilling to rely solely on Soviet gas any longer. They have approached Norway as a more dependable supplier.

Norway, which has large North Sea gas reserves, already supplies EC countries with 12.5 per cent of their gas. It expects this figure to double by the year 2000, when the new Troll gas field comes on stream. It is the world's largest offshore gas field under development.

Algeria in North Africa supplies gas to the EC by two means. Firstly, a trans-Mediterranean pipeline runs to Italy via Tunisia. The second way involves turning gas into a liquid at a temperature of $-161\,°C$, when it takes up to 600 times less space. Liquid Natural Gas (LNG) is shipped to Spain, France and Belgium, where it is once more made into a gas at the terminals.

The world still has large reserves of gas which have not yet been touched. Liquifying it and transporting it by ship could mean that gas will replace oil in importance as oil becomes more and more scarce.

A pipeline is being built to carry Algerian gas from Morocco across to Spain. It will become part of the network of pipelines which link the supply of gas between countries in Europe. The Eastern European countries are keen to be part of this network. The modern high-pressure pipelines can now convey gas with the energy equivalent of 4000 tonnes of coal per hour.

## A NEW USE FOR GAS

A completely new development is the use of gas to produce electricity. Gas has always been seen as too precious for this purpose. However, the efficiency of Combined Cycle Gas Turbines (CCGTs) has reversed this attitude. Natural gas is burnt, and the exhaust gases are used to turn a turbine rather like an aircraft engine. Afterwards, the exhaust gases still have sufficient heat to turn water into steam, and to power a second set of turbines. This form of energy conversion is

**Gas is delivered by lorry to this hotel in Portugal and stored in tanks.**

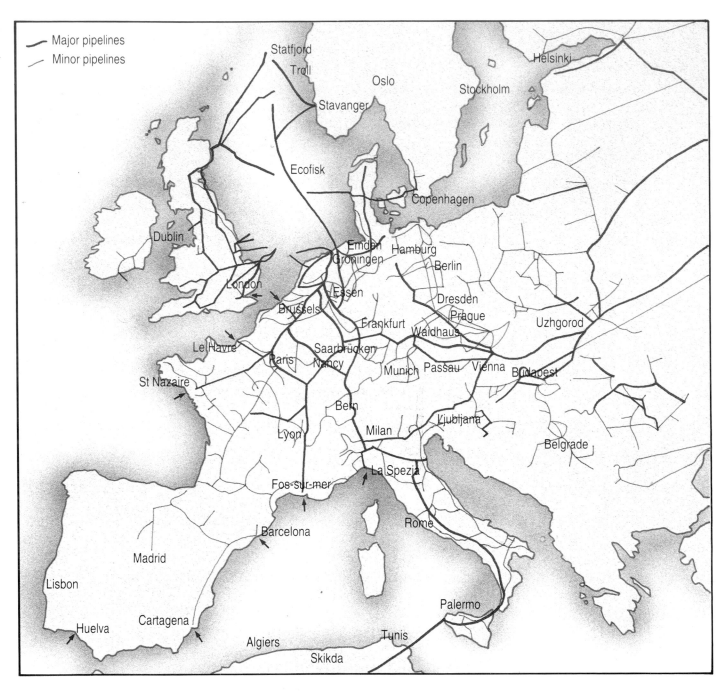

**Europe's gas pipelines.**

considered to be very efficient, much more so than the most modern coal-fired power stations. Because of the optimism about CCGTs, the use of gas in power-stations is expected to increase in Western Europe from 5 per cent in 1988 to 12 per cent in 2005.

# Nuclear power

Nuclear power-stations generate electricity in a similar way to oil- or gas-fired power-stations, except for one crucial difference. Instead of burning a fossil fuel to produce heat, atomic power-stations are fuelled by the energy that is released when atoms of the metal uranium are split apart. After this, the series of energy conversions are the same. Water is heated in a boiler to become steam, this turns the turbines and produces electricity. The electricity is then fed into a country's national grid.

Everything we can see is made from atoms. At the heart of each atom is the nucleus, made up of a combination of two types of particles: protons and neutrons. They are held together by a binding energy. When a uranium atom is hit by another neutron, the nucleus splits, releasing some of its binding energy as heat plus radiation. This leads to a chain reaction, when more and more uranium atoms are split, releasing more and more energy. This process of obtaining energy by splitting atoms is called fission.

Nuclear fission takes place in the core of the reactor. In Europe there are two main kinds of nuclear reactors. Pressurized Water Reactors (PWRs) use pressurized water as a coolant. This takes the heat from the reactor to the boiler. Advanced Gas Reactors (AGRs) use carbon dioxide gas as a coolant.

The uranium is contained in long metal fuel rods in the core of the reactor. An average-sized PWR will contain 50,000 such fuel rods submerged in water. The rate of nuclear fission is controlled by the use of 'moderators' which slow down the speed of the neutrons. In PWRs the water acts as both coolant and moderator. AGRs use graphite rods as moderators to absorb the excess neutrons which are released.

coolants for this kind of reactor.

Apart from fission, there is another kind of nuclear reaction called fusion. This is when two light nuclei join or 'fuse' to make a single, heavier nucleus. The process releases energy, and is similar to the reaction that takes place naturally in the sun. It is still at an experimental stage.

## NUCLEAR SAFETY

Besides heat energy, all nuclear reactions give off radioactivity, invisible rays or waves of energy, released from the nucleus of a radioactive element. Radiation exists naturally in the environment. Artificial radiation is used in medicine, for example in X-ray machines, and in destroying cancer cells. However, strong doses of radiation damage living cells. This means that safety is extremely important in the nuclear power industry. In particular, there is a need for the safe disposal of nuclear waste.

After a time, the uranium fuel elements in a nuclear reactor become used up. They are replaced with new ones, and then reprocessed to take the remaining uranium out. Radioactive waste is left behind, and because the radiation can remain at dangerously high levels for centuries and even thousands of years, it needs to be disposed of somewhere safe.

Left **The German town of Schweinfurth and its nuclear power-station.**

There is an experimental kind of nuclear reactor called the Fast Breeder Reactor (FBR) which uses no moderator at all to control the speed of the reaction. Britain, France and the USSR each have an FBR. The reactor 'breeds' or generates more uranium fuel than it uses, something which sounds impossible. Designing a successful FBR has proved to be a problem, since the extreme heat generated in the core needs to be cooled. Neither water nor gas are suitable

## NUCLEAR POWER IN EUROPE

Concern in Europe about the safety of nuclear power has increased dramatically since 1986 when the world's worst nuclear disaster happened at Chernobyl in the USSR. Clouds of dangerous radioactivity were released over much of the continent.

During the early days of nuclear power in the 1950s and 1960s it was seen as a cheap, safe and reliable means of producing electricity. Since fossil fuels were known to be running out, nuclear power offered countries with limited coal or oil reserves, such as France, Belgium and Hungary, the chance to produce their own energy.

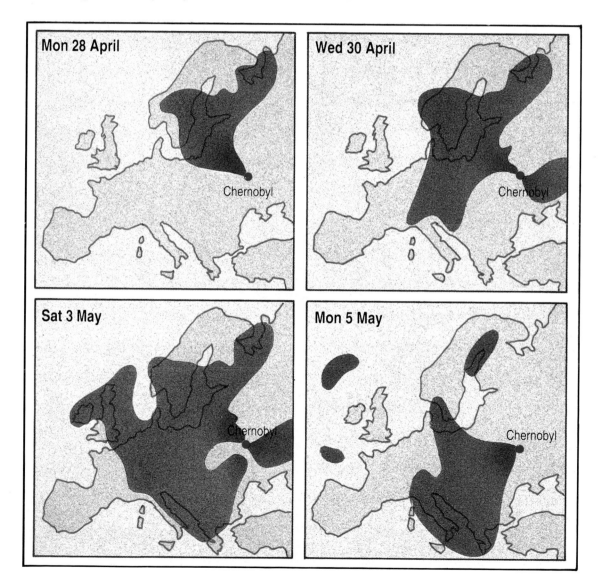

Mon 28 April

Wed 30 April

Sat 3 May

Mon 5 May

Chernobyl

**Areas covered by the cloud of fallout on the days following the nuclear accident at Chernobyl.**

**Balloons released by Greenpeace at the 1991 World Skiing Championships in Austria as a protest against nuclear power.**

## Nuclear energy in France

With 66 per cent of its electricity provided by nuclear power, France relies on this form of energy production more than any other country in Europe. In world terms it is second only to the USA. It has 55 nuclear reactors which provide electricity for the national grid.

With few natural energy resources, France's post-war reconstruction could not have occurred without a reliance on imported oil. Its coalfields in the north-west of the country were uneconomic to run, and while hydroelectric power was developed in the mountainous areas of the Alps and the Pyrenees, it provided only a small proportion of total energy needs.

The oil crises of 1973 and 1979 affected the French economy very severely. The development of nuclear power was seen as a way of ensuring security of energy supply, and so France embarked on its major programme of constructing nuclear plants. Nuclear power is used to supply base load electricity (the average daily needs). This is because nuclear power-stations have to be kept running all the time and they cannot be turned on or off. When demand for electricity increases at peak periods, like at breakfast time, coal- or oil-fired power-stations can be turned on to feed electricity into the national grid.

The French nuclear power programme produces more electricity than is needed, and so France exports the surplus to neighbouring countries such as Switzerland, Italy and Germany. Britain imports the most via a trans-channel cable.

The symbol of French trust in nuclear power until recently has been its Super-phénix fast breeder reactor near Lyons. Started in 1974 and completed in 1986, it has had so many operating difficulties that it is now unclear if the government will continue to support its high costs.

The oil crisis of 1973 made nuclear power seem very attractive. It was seen as an alternative to relying on unstable supplies of oil. France, Belgium and Hungary launched a massive building programme of nuclear power stations. In 1988 they by turn produced 66, 63 and 50 per cent of their electricity by nuclear power. Germany, Spain and Britain also depend on nuclear power, but only half as much as France. The Netherlands and Italy do not produce much nuclear power at all. Many countries simply decided that it was too dangerous. Denmark, Iceland, Ireland, Portugal, Greece and Norway have no nuclear reactors whatsoever.

In 1978 Sweden held a national referendum which decided to phase out its twelve atomic plants by 2010. Austria also voted in 1981 to end its nuclear power programme. Italy voted likewise in 1988 to close its three nuclear plants.

Unless there is a big change in attitude, it looks as if the days of nuclear power are numbered, at least in Western Europe. A combination of high costs and public concern about safety have seen to that. Another accident like Chernobyl would certainly kill it off.

Very few new nuclear power-stations have been commissioned in recent years. Unless new ones are built to replace the ageing power-stations, nuclear power will become less and less important in Europe. However, it is a very difficult problem for countries with little or no natural resources to withdraw from nuclear power altogether. For example, it is unclear how Sweden will fill the gap in electricity production after 2010. Eastern European countries such as Hungary have little choice at present but to continue with nuclear power.

**This pig was born blind as a result of radiation which escaped after the Chernobyl disaster of 1986.**

# Energy conservation

It is a well-known fact that the world's reserves of fossil fuels are running out. The oil crises since 1973 have prompted both greater use of gas and serious research into alternative, renewable sources of energy, such as solar power. They are explored in the following pages. A second effect of the oil crises has been to find ways of simply using less energy. One way of saving energy is by using it more efficiently. It is easier to conserve one unit of energy than it is to generate a new one. Environmentalists tell us that we cannot go on using energy as freely as we do; the pollution caused by excessive energy consumption is too harmful to the environment.

The energy used to provide heat and light for buildings (domestic, public and industrial) in Britain accounts for an astonishing 45 per cent of all the primary energy used. Simple conservation techniques could reduce this figure by half, and the buildings would still be just as warm. It is all a question of efficiency.

**Energy can be conserved by lagging floors with glass-wool. This prevents warm air escaping upwards from the room below.**

Insulation is one answer: cavity wall insulation, loft insulation, double-glazing and draught excluders all help reduce heating costs. Insulation is cheap and works by trapping air, stopping it from carrying the warmth away.

In Scandinavia, where winter is longer and far colder than in Britain, houses are extremely well insulated. Windows, for example, are not just double-glazed but often triple-glazed.

Another form of energy conservation that is well-developed in the Scandinavian countries, Eastern Europe and Germany comes from Combined Heat and Power (CHP) stations. These stations make use of the heat that is left in steam or water after it has been used to generate electricity.  Very often, the heat is wasted when released up the chimney or into a river, but in these countries it is piped away, and provides space heating for homes and industry. The efficiency of the system can be as high as 70 per cent, since it doubles the energy output from each unit of primary energy.

It is often said that we live in a 'throwaway society'. Since conservation is all about reducing the amount of energy we waste, new ways have been found to use the energy stored in household refuse. Recycling glass bottles, aluminium cans and newspapers helps to conserve energy, but it is only a beginning.

Before long more countries will see the reasons for following Sweden and Germany's example of sorting and grading household waste into different categories. The organic waste (vegetable peelings, paper, cardboard) can then be made into pellets which burn cleanly at high temperatures. The heat from municipal waste burners and factories provides district heating when it is distributed by water pipes to city housing areas.

**Europe's environment can be kept cleaner by using oil reclamation plants such as this one.**

# Renewable energy

Fossil fuels are non-renewable forms of energy. They took millions of years to form, so once they are gone they are gone forever. Indirectly, they store in chemical form the solar energy that once made those prehistoric forests grow. So why not turn directly to the sun as an energy source? After all, life on earth would be impossible without it. And it is there every day, even when hidden by cloud! Different natural energy flows, such as wind, waves and rivers, are also sources of renewable energy.

These energy forms occur because of the sun. Winds blow as a result of air movements caused by the sun heating the atmosphere. The waves of the sea are in turn caused by wind, while tides result from the gravitational pull of the sun and moon. The sun's heat is also responsible for the water cycle, which causes water to evaporate from seas and to fall on high ground and feed rivers. Each of these forms of renewable energy can be harnessed. Some advantages of renewable energy are:

- Renewable energy like solar and wind power does not have to be imported. It is 'home-grown'.
- Its diversity offers flexibility of energy supply. This is important when trying to reduce dependency on fossil fuels.
- It does not contribute to global warming.

Some disadvantages are:

- Many renewable sources of energy are not available all the time. For example, there is least sun in winter, when energy is most needed.
- Lack of research into the renewables makes the generators expensive to produce in small quantities.
- Renewable energy does not come in as concentrated a form as fossil fuel or nuclear energy.

Left **As our reserves of fossil fuels run out, we will have to make greater use of renewable sources, such as solar power.**

## SOLAR POWER

Only one thousandth of a millionth of the sun's energy output is received by earth, and 30 per cent of this is reflected back into space by the ozone layer in the upper atmosphere. Even so, the amount that reaches us is more than enough to provide for all our needs. The sunlight that falls on the earth each year is 10,000 times more than the total amount of energy used by everybody on earth.

Wind, wave and hydroelectric power all generate electricity. Solar power can not only do this, but can also be used to heat water and buildings.

Active solar power uses equipment or machinery that moves. One form converts sunlight into electricity by using photovoltaic cells, which were developed for the space programme. They are very useful for providing power in remote places that do not have mains electricity. Although expensive, production costs are falling. This type of solar power is already used to provide energy for some villages in Spain. Solar cells are also used to power calculators and watches.

A solar furnace at Odeilo in France concentrates the sun's rays on to a large mirror, which in turn focuses the light on to a furnace. Temperatures of 3,300° C have been reached.

Active solar heating systems use solar panels or collectors placed on a south-facing roof to get heat from the sun. A liquid such as oil or water carries the heat away, and is stored until needed. Like solar cells, the technology for solar heating is still expensive, but prices will fall when demand for it grows. In Germany, large public buildings such as hospitals are already turning to solar power because of its long-term cheapness.

Although active solar power would seem to be of more use in the warmer Mediterranean countries, an experiment in Sweden shows that it can be used to provide space-and-water heating. Generated in summer, the heat is stored in an underground rock cavern until it is needed in winter.

Passive solar power relies on the simple fact that in Europe it is always sunnier facing south rather than north. Buildings can be designed with large south-facing

**An experimental solar power-station at Almeria in Spain.**

windows to trap the sun's heat. If this heat is then prevented from escaping by good insulation, fuel bills can be halved.

## WIND POWER

Windmills have been used in Europe to grind corn into flour for many centuries. During the seventeenth and eighteenth centuries they were used as water pumps to assist with land drainage in Holland and East Anglia. Wind power was first used to generate electricity in the 1860s. However, it was the oil crises of the 1970s that led to the development of large-scale wind turbines or 'aerogenerators' as a renewable energy source capable of producing electricity.

There are two sorts of windmill. Traditional ones have a horizontal axis which needs to face into the wind. Then there is the more modern design with a vertical axis which does not need to face the wind. The larger the area covered by the blades, the greater the power output. However, whatever the design, windmills always lose a lot of power to friction. Even the best design will only obtain about 45 per cent of the energy available in the wind.

Since the strongest and most reliable winds occur along coastlines, there are many European countries which could benefit from using wind power. The west coast of Britain is particularly well placed to make use of the strong Atlantic winds. What is more, winds blow harder in winter, when we need more electricity for warmth and light.

Unfortunately, wind power has not been taken as seriously in Britain as it has in Denmark. Investment has been lacking. As a result there are only a few experimental sites, and plans to build just three wind farms in Cornwall, Wales and Scotland. Nonetheless, it is estimated that Britain could generate half of its electricity from the wind. But producing more than 20 per cent of electricity by wind power would require expensive back-up from conventional oil- or coal-fired power-stations to cover the supply on calm, windless days.

The plans for the Welsh wind farm were opposed in autumn 1990 by countryside groups which said the 25 m high turbines would spoil the view. In the near future, choices are going to have to be made between such conflicting issues as the siting of wind farms, electricity demand and the need for a cleaner environment. Such problems have led the Danish to site wind farms offshore in shallow coastal waters.

Denmark already has three large wind farms and numerous small wind turbines which provide 1 per cent of the country's electricity. It plans to increase this figure to 10 per cent by the year 2000. The Netherlands and Spain have also constructed wind farms, whilst Germany and Sweden are developing major wind power programmes.

## WATER POWER

The ancient Greeks were probably the first to make use of the power of moving water. They used water wheels to grind corn. Water wheels have been used for the same purpose throughout Europe right until this century. Today there are three forms of water power: hydroelectric (HEP), tidal and wave power.

HEP transforms the potential energy stored in lakes into kinetic energy when water passes through a pipe at great pressure, and turns turbines which then drive electrical generators. The 'head' is the distance of the lake above the turbines. The greater the head, the more power that is generated.

Once a valley is dammed and the turbines are constructed, HEP provides the cheapest form of electricity. Before they turned more to imported oil for their energy supplies, northern European countries such as Sweden and Finland and southern European countries such as Portugal, Italy and Greece relied heavily on HEP. Norway still generates 90 per cent of its electricity by HEP, although this figure is falling rapidly because of the increased use of their North Sea oil and gas reserves. Switzerland still generates 60 per cent of its electricity by means of HEP.

Tidal power exists more as an idea than as practical reality in

**A dam in the Austrian Alps has created a large lake. The water is used to generate hydroelectric power.**

## Energy in Sweden

Sweden did not require large amounts of energy until this century, when it began to industrialize and the population began to move from the countryside to the cities. Hydroelectric power was developed, but it was the use of imported oil which allowed rapid industrialization to occur.

In 1970, oil accounted for 70 per cent of the country's energy supply, but the oil crises led to a wider use of other energy sources such as coal and nuclear power. By 1988 reliance on oil had been cut to less than half of the energy supply. Hydroelectric and nuclear power account for one-third of the energy supply while the wood processing industry provides wood fuels from logs, chips, bark and sawdust.

In 1978 Sweden held a national referendum which decided to phase out its twelve nuclear power-stations by 2010.

This has led to a lot of debate about how the country will restructure its energy policy to ensure electricity supply.

Equal quantities of electricity are produced by hydroelectric and nuclear power. Use of coal- and oil-fired power-stations will have to increase. However, Sweden also has high standards of environmental protection, particularly because of the destruction of its lakes and forests caused by acid rain.

Sweden has drastically reduced the sulphur dioxide released from its own coal- and oil-fired power-stations. Since pollution crosses national boundaries, it has also played a large role in pressing for greater pollution controls throughout Europe.

It is likely that an increased use of natural gas will have an important part to play while the country phases out nuclear power. At present, it buys gas from both Norway and the USSR.

Europe. It makes use of the twice-daily rise and fall, or ebb and flow, of the tides. There was a tidal mill on London Bridge in the seventeenth century. The modern show-case example is at the mouth of the River Rance, near St Malo in France. It was built in 1968 and obtains 92 per cent of the power available from the tidal flow at that point. Another has been built at Murmansk in the USSR.

There has been talk for more than fifty years of building a tidal barrage in the Bristol Channel. The large tidal range makes it an attractive site for harnessing tidal power. Some estimates say that it could produce 7 per cent of Britain's electricity needs. One thing is sure, the longer the project is left, the more expensive it will be to build.

Wave power is still at the experimental stage, and the many different designs which exist prove that it would be an important source of energy, if only sufficient investment was put into it.

One design, Salter's Duck, rocks

up and down in the waves. The movement compresses the air inside, which transfers the wave energy to a turbine.

## GEOTHERMAL ENERGY

The word 'geothermal' means heat from rocks. Molten rock, or lava, which bursts from volcanoes gives us an idea of how much heat there is within the earth. The heat derives from the decomposition of radioactive substances which have been present at the earth's core ever since it was formed. Temperatures there are estimated to be in excess of 4,500° C. Strictly speaking, geothermal energy is not a renewable source of energy at all, because the earth's heat is finite.

Geothermal energy is harnessed in places of volcanic activity where the earth's crust is thinner than at more stable places. In Europe this means Iceland, with its volcanoes, geysers and hot springs. Reykjavik, the capital, has all its domestic and industrial heating and hot water provided for by this means. Italy also makes use of its region of volcanic activity, around Lardarello. Steam from below the earth's crust is used to generate electricity.

A recent development is called Hot Dry Rock technology. Two holes are drilled 2-3 km down through rock, where temperatures can be high enough to boil water.

Explosives are used to break the rock up, and water is then pumped down the first well at high pressure. The water is heated up by the hot rock, and returned under pressure up the second well. The hot water can be used for generating electricity or district heating.

**All of the heating required by Iceland's capital, Reykjavik, is provided by geothermal energy.**

# The way forward

The end of the Cold War between East and West has brought many changes to Europe. It will continue to do so. The European Community is likely to grow in size and importance, bringing increased co-operation in the field of energy policy. One idea put forward by the Netherlands is for the founding of a European Energy Community to co-ordinate energy policy.

Such a Community would bring together East and West at the end of the Cold War, just as the ECSC succeeded in reconciling the old enemies of France and Germany after the Second World War.

Problems to do with energy will not vanish by themselves. Even a unified Europe will still have to further reduce its dependency on oil, as reserves continue to run out. It is likely that interest in the renewable sources of energy will increase, while energy conservation will have to be improved, particularly in Eastern Europe. New developments in pollution control mean that coal still has a major part to play in European energy policy. Gas will become increasingly important. The safety aspects of nuclear power give this form of energy an uncertain future in Europe.

Europe must also continue to pay attention to the quality and health of the environment. Environmental pollution does not respect national boundaries. The widespread effects of acid rain and the greenhouse effect, resulting from energy production, affect the whole world. Global problems will require joint international action, reminding us that the continent of Europe is part of the rest of the world.

Left **Energy for the future: renewable sources such as wind power will become increasingly important.**

Opposite page **The formation of the EC.**

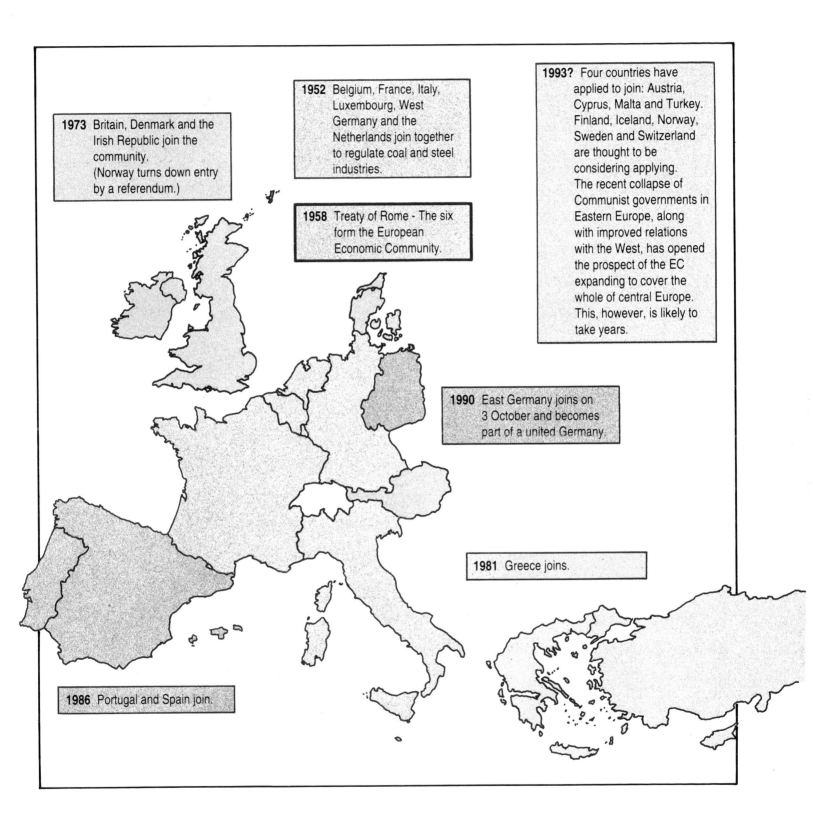

1952 Belgium, France, Italy, Luxembourg, West Germany and the Netherlands join together to regulate coal and steel industries.

1958 Treaty of Rome - The six form the European Economic Community.

1973 Britain, Denmark and the Irish Republic join the community.
(Norway turns down entry by a referendum.)

1993? Four countries have applied to join: Austria, Cyprus, Malta and Turkey. Finland, Iceland, Norway, Sweden and Switzerland are thought to be considering applying. The recent collapse of Communist governments in Eastern Europe, along with improved relations with the West, has opened the prospect of the EC expanding to cover the whole of central Europe. This, however, is likely to take years.

1990 East Germany joins on 3 October and becomes part of a united Germany.

1981 Greece joins.

1986 Portugal and Spain join.

# Glossary

ACID RAIN  Rain, snow and mist that has absorbed pollution in the atmosphere and become acidic.

CALORIFIC VALUE  The unit used to measure the quantity of heat released by a fuel.

CAPITALISM  An economic theory which states that private wealth or 'capital' is used in the production and distribution of goods in order to generate profit.

COLLIERIES  Coal-mines.

COMBINED CYCLE GAS TURBINES (CCGTS)  A new and highly efficient method of generating electricity from burning natural gas.

COUNCIL FOR MUTUAL ECONOMIC ASSISTANCE (COMECON) The trading organization of the former socialist Eastern European countries founded in 1949.

EASTERN BLOC COUNTRIES  The Eastern European countries such as Romania, Hungary and Czechoslovakia, which until recently were under Communist government.

EUROPEAN ATOMIC ENERGY COMMUNITY (EURATOM) Set up in 1957 to encourage co-operation over the safe development of nuclear power between EC countries.

EUROPEAN COAL AND STEEL COMMUNITY (ECSC) A forerunner of the EC, founded in 1951 to co-ordinate energy policy between 'the Six'.

EUROPEAN COMMUNITY (EC)  Consists of Belgium, France, Italy, Luxembourg, Germany, the Netherlands, Britain, Denmark, the Irish Republic, Greece, Portugal and Spain (see page 45).

FAST BREEDER REACTOR (FBR)  An experimental nuclear reactor which uses no moderator to control the speed of the reaction.

FLUE GAS DESULPHURIZATION (FGD)  Filters which remove sulphur dioxide pollution from the chimneys of coal-fired power-stations.

GREENHOUSE EFFECT  The increase in the average temperature of the earth caused by the build-up of pollutants in the atmosphere.

HYDROCARBONS  Complex compounds of hydrogen and carbon which occur in forms like natural gas and crude oil.

KINETIC ENERGY  The energy of a moving object.

LIGNITE  A low-grade coal with low carbon and high sulphur content.  Also called brown coal.

LIQUIFIED NATURAL GAS (LNG)  Natural gas which has become a liquid after being cooled to $-161°$ C.

MODERATOR  A substance used to control the rate of nuclear fission by absorbing neutrons.

NEUTRONS  Particles which begin the process of nuclear fission.

NUCLEAR FISSION  The process by which energy is released when uranium atoms are split by neutrons.

NUCLEAR FUSION  The process by which energy is released when two light nuclei join to make a single, heavier nucleus.

NUCLEUS  The main mass of an atom.

PHOTOVOLTAIC CELLS  Used to generate electricity from solar energy.

RENEWABLE ENERGY  Generated by using energy from the sun, wind or water.

SINGLE ENERGY MARKET  One energy market for all of the EC.

SINGLE MARKET  The EC's policy which from 1993 will encourage a free market between all member countries.

SOCIALISM  An economic theory where the state should own all important means of the production and distribution of wealth.

SPACE HEATING  Heat used to provide warmth for buildings.

# Books to read

Clark, Ronald W: The Greatest Power on Earth (Sidgwick and Jackson, 1980)

Crisp, Tony: Energy (Nelson, 1983)

Driscoll, Vivienne: Focus on Nuclear Fuel (Wayland, 1985)

Flood, Michael: Energy Without End: The Case for Renewable Energy (Friends of the Earth, 1986)

Gibson, Michael: The Energy Crisis (Wayland, 1987)

Hawkes, Nigel: Nuclear Power (Wayland, 1989)

Jones, John: Energy (Blackwell, 1989)

Lambert, Mark: Focus on Oil (Wayland, 1986)

Lambert, Mark: Focus on Radioactivity (Wayland, 1989)

Lambert, Mark: Future Sources of Energy (Wayland, 1986)

Neal, Philip: Energy, Power Sources and Electricity (Batsford, 1989)

Pechey, Roger: Focus on Gas (Wayland, 1986)

Ramage, Janet: Energy: a Guidebook (OUP, 1983)

Rickard, Graham: Solar Energy (Wayland, 1990)

Rickard, Graham: Water Energy (Wayland, 1990)

Ross, Stewart: Towards European Unity (Wayland, 1989)

Rowland-Entwistle, Theodore: Focus on Coal (Wayland, 1987)

Spurgeon, Richard, and Flood, Mike: Energy and Power (Usborne, Science and Experiments series, 1990)

# Further information

You can contact these organizations to find out more about the issues covered in this book.

### COAL
British Coal – Schools Service
Public Relations Department
Room 457 Hobart House
Grosvenor Place
London SW1X 7AE

### ELECTRICITY
Understanding Electricity
PO Box 44
Wetherby
W. Yorkshire LS23 7ES

### NUCLEAR POWER
British Nuclear Fuels plc
Information Services
Sellafield
Seascale
Cumbria CA20 1BR

### OIL AND GAS
British Gas Education Service
PO Box 46
Hounslow
Middlesex TW4 6NF

BP Educational Service
Britannic House
Moor Lane
London EC2Y 9BU

### ALTERNATIVE ENERGY
CEAT (Coordination Européenne des Amis de la Terre)
Rue Blanche, 29
1050 Brussels
Belgium

Friends of the Earth
26-28 Underwood Street
London N1 7JQ

# Index

acid rain 24, 44
advanced gas reactors 31
Austria 6, 7, 35, 41

Baltic republics 6
Belgium 11, 12, 13, 14, 15, 21, 23, 29, 33, 35
Britain 12, 13, 14, 19, 20, 21, 23, 25, 26, 32, 34, 35, 37, 40, 42
Bulgaria 7

Chernobyl 33, 35
Churchill, Winston 7, 12
coal 8, 9, 10, 12, 14, 15, 19, 21–5, 33
    coal-fired power-stations 30
    community 11
    fields 12,34
    mines 13, 22, 23
Cold War 7, 18, 44
Combined Heat and Power stations 37
Council for Mutual Economic Assistance 16, 18
Cyprus 7
Czechoslovakia 7, 17, 24, 28

Denmark 13, 19, 20, 35, 40

Eastern Europe 7, 16, 17, 18, 24, 26, 28, 35, 37
economy 18
electricity 9, 10, 29, 33, 34, 41, 43
European Atomic Energy Community 11
European Coal and Steel Community 11
European Community 6, 7, 10, 11, 13, 14, 15, 17, 18, 25, 27, 29, 44

Fast Breeder Reactors 32
Finland 6, 7, 23, 41
fossil fuels 6, 24, 33
France 11, 12, 13, 14, 15, 21, 23, 29, 32, 33, 34, 35, 39, 42

gas 4, 6, 8, 9, 10, 14, 18, 19, 20, 24, 28, 29, 30, 31, 42, 44
    fields 20
    gas-fired power-stations 9
    liquid natural gas 29
geothermal energy 43
Germany 7, 11, 12, 13, 18, 19, 20, 21, 23, 24, 31
Gorbachev, Mikhail 18
Greece 13, 35, 41
greenhouse effect 24, 44

Holland 40
Hungary 7, 33, 35
hydroelectric power 9, 23, 34, 39, 41, 42

Iceland 7, 35, 43
Industrial Revolution 21
Ireland 13, 35
Italy 11, 13, 14, 29, 34, 35, 41

Lithuania 16
Luxembourg 11,13,14

Malta 7
Marshall Plan 7
Mediterranean 39
miners 22, 23

national grid 34
Netherlands, the 11, 12, 13, 14, 15, 19, 20, 23, 27, 35, 41
North Sea 14, 19, 20, 29, 41
Norway 7, 14, 20, 28, 29, 35, 42
Nuclear power 8, 9, 10, 14, 31–5

oil 8, 9, 10, 12, 13, 17, 19, 20, 24, 26–7, 31, 33
    crises 14, 27, 35, 36, 40, 41, 42
    prices 13
ozone layer 39

petrol 6, 9
Piper Alpha 20
Poland 7, 15, 17, 19, 28
pollution 24, 25, 44
Portugal 13, 14, 23, 29, 35, 41
power-stations 35
Pressurized Water Reactors 31

recycling 37
renewable energy 8, 14, 24, 36, 37, 38
Romania 16, 17
Rotterdam 12, 13

Scandinavia 6, 24, 37
Second World War 6, 7, 12, 23
Single Energy Market 14, 15
solar power 8, 14, 24, 36, 38, 39, 40
Spaak, Paul Henri 11
Spain 12, 13, 23, 29, 35, 39, 41
Sweden 6, 7, 23, 35, 37, 39, 41, 42

USA 6, 7, 15, 23
USSR 6, 7, 16, 17, 18, 19, 32, 33, 42

Wales 22
Watt, Isaac 21
wave power 9, 39, 41, 42, 43
Western Europe 11, 12, 13, 15, 16, 18, 19, 21, 23, 26, 28, 30
wind power 8, 24, 39, 40, 41